THIS BOOK BELONGS TO:

....

LEARNING NUMBERS

Email us at
tgosketch@gmail.com
ISBN 978-1-955126-34-2

TgosketchPress
Chicago, Illinois
www.tgosketch.com

Hi, friend! It's me Marsee!
Did you know numbers aren't just for counting things like snacks or socks?
Numbers can also tell stories about feelings, friendship, and all the little moments that help us grow.

Every number means something special.
Like how 1 can remind us of a big heart…
Or how 10 can be the steps we take when we're feeling brave.
Even 20 can be the ways we spread kindness in the world!

You don't have to be perfect just curious.
So take your time, count with me, and feel each number as we go.

Just remember:
Look at the number
Think about what it means
Let it remind you how amazing you are
We'll learn together, one number at a time.
And every time you count, you're growing stronger, kinder, and more you.
Let's go find some numbers, okay?

Love, Marsee

1 LEARNING NUMBERS

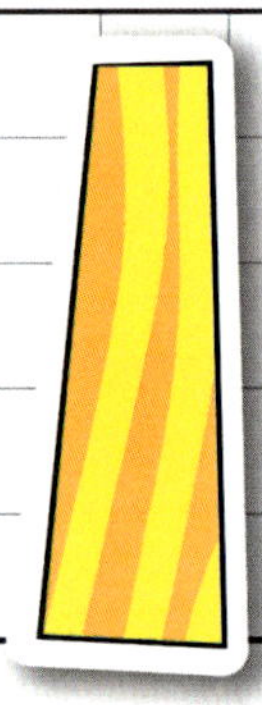

COLOR THE NUMBER ONE

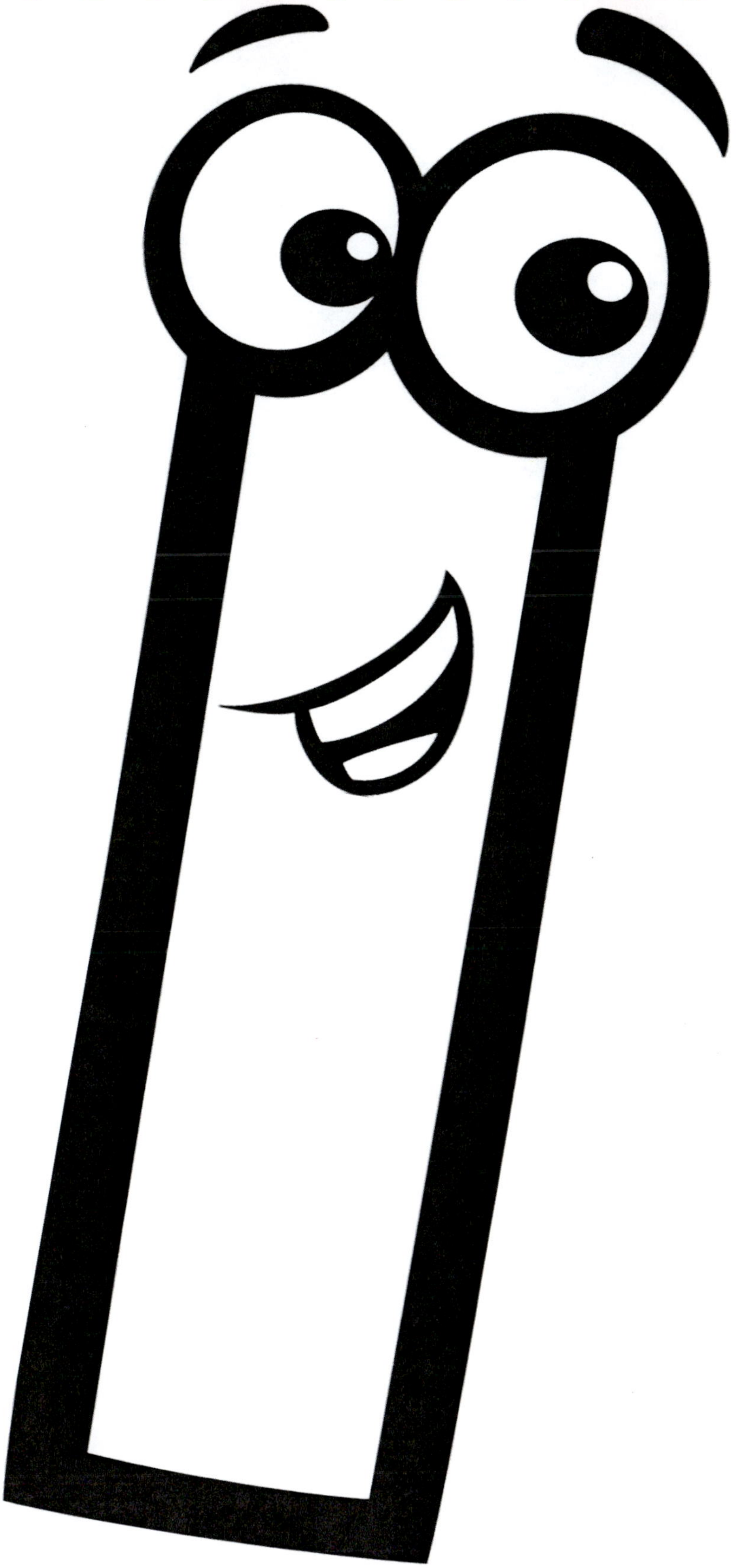

2 LEARNING NUMBERS

2

2 2 2 2 2 2 2 2 2 2

2 2 2 2 2 2 2 2 2 2

2 2 2 2 2 2 2 2 2 2

2 2 2 2 2 2 2 2 2 2

2 2 2 2 2 2 2 2 2 2

COLOR THE NUMBER TWO

3 LEARNING NUMBERS

3

3 3 3 3 3 3 3 3 3 3

3 3 3 3 3 3 3 3 3 3

3 3 3 3 3 3 3 3 3 3

3 3 3 3 3 3 3 3 3 3

3 3 3 3 3 3 3 3 3 3

COLOR THE NUMBER THREE

COLOR THE NUMBER FOUR

4 LEARNING NUMBERS

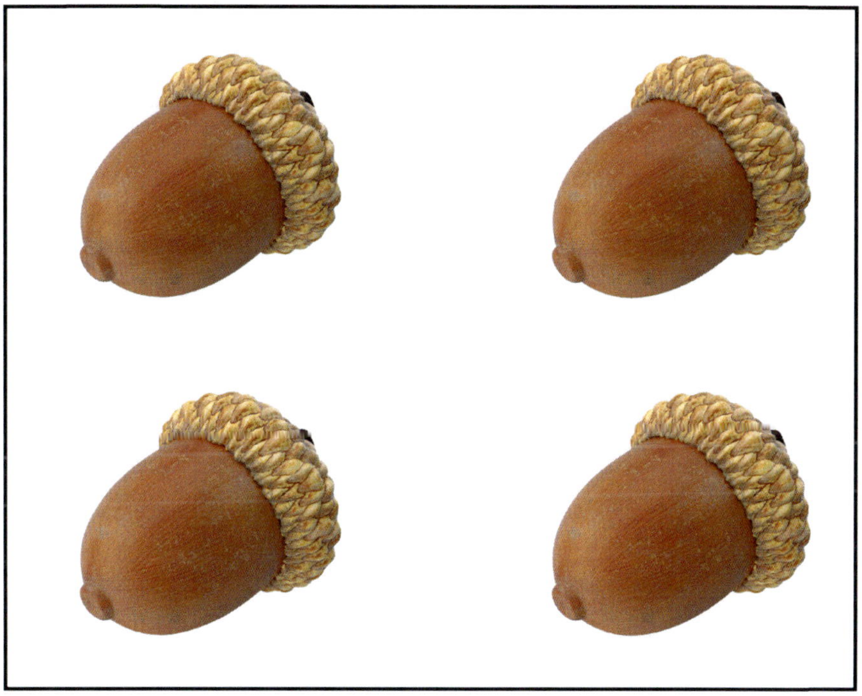

4

4 4 4 4 4 4 4 4 4 4

4 4 4 4 4 4 4 4 4 4

4 4 4 4 4 4 4 4 4 4

4 4 4 4 4 4 4 4 4 4

4 4 4 4 4 4 4 4 4 4

COLOR THE NUMBER FIVE

5 LEARNING NUMBERS

5

5 5 5 5 5 5 5 5 5 5

5 5 5 5 5 5 5 5 5 5

5 5 5 5 5 5 5 5 5 5

5 5 5 5 5 5 5 5 5 5

5 5 5 5 5 5 5 5 5 5

COLOR THE NUMBER SIX

6 LEARNING NUMBERS

6

6 6 6 6 6 6 6 6 6 6

6 6 6 6 6 6 6 6 6 6

6 6 6 6 6 6 6 6 6 6

6 6 6 6 6 6 6 6 6 6

6 6 6 6 6 6 6 6 6 6

COLOR THE NUMBER SEVEN

7 LEARNING NUMBERS

7

7

7

7

7

7

8 LEARNING NUMBERS

8

8

8

8

8

8

COLOR THE NUMBER EIGHT

9 LEARNING NUMBERS

9

9
9
9
9
9

COLOR THE NUMBER NINE

10 LEARNING NUMBERS

10

10 10 10 10 10 10 10 10

10 10 10 10 10 10 10 10

10 10 10 10 10 10 10 10

10 10 10 10 10 10 10 10

10 10 10 10 10 10 10 10

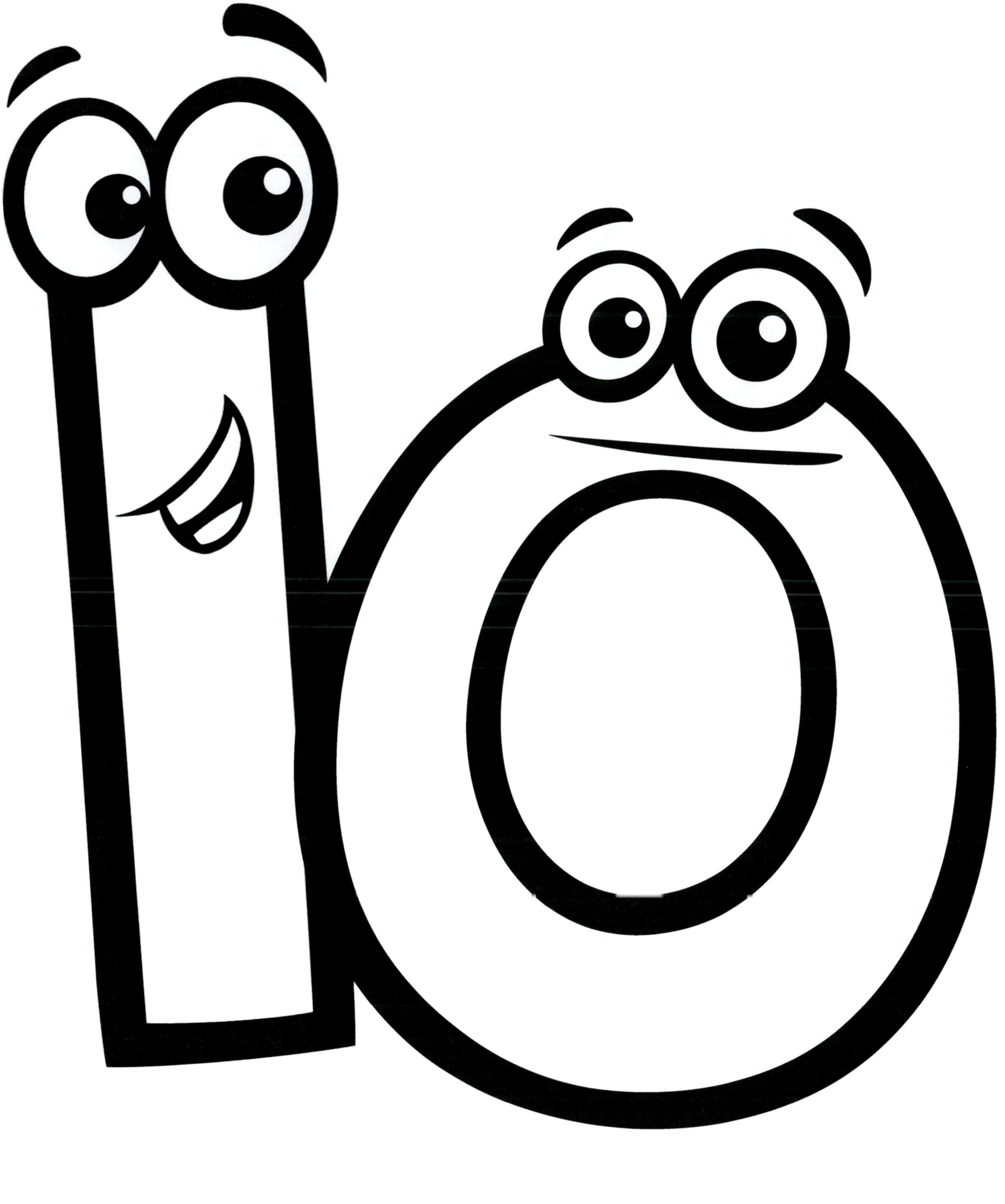

COLOR THE NUMBER TEN

LET'S WRITE OUR NUMBERS

1 1 1 1 1 1 1 1 1 1

2 2 2 2 2 2 2 2 2 2

3 3 3 3 3 3 3 3 3 3

4 4 4 4 4 4 4 4 4 4

5 5 5 5 5 5 5 5 5 5

6 6 6 6 6 6 6 6 6 6

7 7 7 7 7 7 7 7 7 7

8 8 8 8 8 8 8 8 8 8

9 9 9 9 9 9 9 9 9 9

10 10 10 10 10 10 10 10

LET'S WRITE OUR NUMBERS

1 1 1 1 1 1 1 1 1 1

2 2 2 2 2 2 2 2 2 2

3 3 3 3 3 3 3 3 3 3

4 4 4 4 4 4 4 4 4 4

5 5 5 5 5 5 5 5 5 5

6 6 6 6 6 6 6 6 6 6

7 7 7 7 7 7 7 7 7 7

8 8 8 8 8 8 8 8 8 8

9 9 9 9 9 9 9 9 9 9

10 10 10 10 10 10 10 10